Strategic Information in Note Form for Veterinary Students and Others
Peter Curtis

First published by the
Department of Veterinary Clinical Science
University of Liverpool

Second edition published 1987 by
Liverpool University Press
Senate House, Abercromby Square, Liverpool, L69 3BX

Third edition 1990 as A Handbook of Poultry and Game Bird Diseases

Fourth edition 1996

British Library Cataloguing in Publication Data
A British Library CIP record is available

ISBN 0-85323-610-0

Printed in the European Union by
The Alden Press in the City of Oxford

Contents

Preface

The veterinary undergraduate course is full of information on, and instruction in, various disciplines relevant to poultry problems, for the earlier parts of the course anatomy and physiology, biochemistry and husbandry, nutrition, pharmacology, and genetics, and other disciplines are covered - later pathology, meat hygiene, parasitology, and microbiology are taught, and the result can be a mind which is very well informed factually, but liable to be confused by the interaction of these various disciplines when contemplating clinical problems on poultry farms.

These notes represent a continuing attempt to provide veterinary students and others with an 'abbreviated overall view of the current disease situation' as was stated in the preface to the Second Edition in 1987. Obviously, more specialised sources will have to be consulted for specific problems, and these may be textbooks, veterinary journals, poultry practitioners, or consultants in various fields. Poultry problems are frequently urgent and the clients likely to be demanding, particularly when large numbers of valuable stock are at risk. Telephone consultations with colleagues are a feature of veterinary poultry work, especially when new or unfamiliar conditions are appearing.

In the 1987 Preface it was suggested that absence of serious disease had led UK farmers to be "less cautious" about disease but by the third editions appearance (1990) Salmonella enteritidis and new forms of Infectious Bursal Disease had cured any complacency. Now we have concern that more virulent Marek's Disease is present and earlier introductions such as Turkey Rhinotracheitis (1985) and Pigeon Paramyxovrus 1 (1983) are well established. The implications of the more liberal ("open frontier") EU policy for imports from Europe are unclear at present. Clearly excessive reliance on vaccines is unwise. Experience with infectious disease problems in poultry populations is often a useful indicator for trends in other species.

An environmental approach to poultry problems is strongly recommended to the clinician invited to investigate a disease episode, since such episodes are frequently multifactorial in origin. Veterinary training is extremely multidisciplinary and an excellent grounding for such an approach. Welfare considerations are increasingly of concern to the consumer, and clinicians must help and advise their clients on the achievement and maintenance of high welfare standards.

Poultry and birds are intrinsically interesting and we have much yet to learn of their biology and of their diseases. Veterinary interest in birds and poultry is growing and will continue to grow, resulting in greater knowledge and greater understanding of their needs.

Acknowledgements are due to Ian Cameron for his contribution, to poultry and game farmers for introducing me to some of their problems, and to University colleagues for helping to find some answers. Editorial guidance was provided by John Cox and Robin Bloxsidge. The opinions expressed in this book are personal, as is the responsibility for any errors or omissions.

PETER CURTIS

Investigation, Diagnosis and 'Trouble-Shooting'

Poultry farmers and others require help when early signs of disease or poor performance are suspected in their flocks, and with the reduced availability of free advice in Britain, there is an increasing opportunity for local veterinary practitioners to undertake basic 'trouble-shooting' investigations into these problems, many of which initially require a clinical and environmental medicine approach.

Poultry practitioners with specialised laboratory facilities, may advertise their services, offering a comprehensive advisory service on health, productivity, husbandry, environmental problems, welfare, incubation problems, and hygiene, both on the farm and in processing.

A local practitioner can collaborate with such a practice to investigate problems occurring in his or her district, possibly many miles from a laboratory. The local person has the advantage of familiarity with the site, its staff and its standards, and can provide a quick 'on-the-spot' assessment and at least a provisional diagnosis. If necessary appropriate specimens can then be selected for further tests in conjunction with the laboratory. Medication can be rapidly prescribed if necessary, but in many cases improved husbandry may resolve the problem.

Diagnosis is concerned with identifying the nature of a disease but it is also of great help to worried farmers to have a 'negative' diagnosis as when the veterinarian can exclude certain worrying possibilities.

A   Clinical Diagnosis and Environmental Medicine
This relates to the bird, and all the factors impinging on the bird, in its normal environment, and is an essential first step in any investigation. The environmental medicine approach will make use of the veterinarian's multidisciplinary skills and will include consideration of such factors as climate, behaviour (animal and human), welfare, genetic

make-up, drug interactions, diet, housing, vaccinal history, economic realities, parasitic and microbiological agents, and such others as appear relevant. These aspects have to be integrated with the examination of individual normal and abnormal birds and study of a detailed history of the problem.

B   Laboratory Diagnosis

Traditionally much poultry disease diagnosis has been based on the examination of live or dead birds despatched to laboratories for examination, autopsy and other tests as were appropriate. In the past specific infectious diseases were relatively more important and prevalent, e.g. fowl typhoid, avian tuberculosis, than today, when more complex interactions are likely to be involved. Sometimes carcases from one outbreak or problem finish up at several laboratories despatched by different interested parties - farmer, husbandry advisor, food compounder - this happens when there is a dispute about the nature of the problem.

Laboratory-based vets insist on a good history and may phone the farm to correct deficiencies which may be vital, e.g. the precise date when symptoms first noted.

Food samples are sent to analytical laboratories to check for nutritional contents, anti-coccidial drug level, mycotoxins, etc.

Some Diagnostic Techniques

Bacterial Disease
Bacterial culture: e.g. fowl cholera, salmonellosis etc.
Toxin detection, e.g. botulism
Serology - e.g. pullorum disease and invasive salmonellae. mycoplasmosis
Impression smears: chlamydia, tuberculosis, other bacteria (erysipelas, staphylococcus, pasteurella, etc.)
Fluorescent antibody: clostridia

Parasitic Disease

Microscopic examination for protozoans, worms' eggs, mites, etc.

Viral Disease

1 Virus isolation techniques: tissue culture

                                          egg inoculation

2 Electron microscopy: useful for pox and infectious larynogo-tracheitis (ILT).

3 Serology: widely used for (a) screening healthy flocks to ensure that the vaccine administered has had the desired effect. Sometimes live vaccines have been omitted, or killed by chlorine in water resulting in lack of protection; (b) aiding diagnosis as when a raised titre accompanies an outbreak of disease or a performance 'hitch' implying that the two are connected. There is much discussion then as to the significance of any 'sero-conversion' which has to be related to other factors on the farm.

Serological tests used include plate tests, agar gel. haemagglutination inhibition, serum neutralisation and ELISA. Serological tests have been bedevilled by erratic antigens but a series of ELISA techniques are being marketed by several organisations.

C   Epidemiological Diagnosis

Since poultry tend to live in large flocks the consideration of disease patterns in a flock as for example in acute runting syndrome, may enable a diagnosis to be at least partially based on these observations When a flock has completed its life it is often advantageous to analyse what problems occurred and thus avoid a repetition in the next flock.

D   Autopsies

The term autopsy (seeing with one's own eye) is used for the examination of carcases and this is a vital activity for the clinician concerned with poultry since it is an essential part of virtually all investigations.

General comments

1   Autopsies may be undertaken for
    a   General screening check on the health of a flock.
    b   Diagnosis of a well defined problem.
    c   Other reasons e.g. to exclude a particular diagnosis.

2   Negative findings can be very valuable to the poultry farmer.

3   In rural practice autopsies on the farm may be valuable if the laboratory is some long distance away.

4   A full flock history is desirable before the autopsy, at a laboratory, begins. When the autopsy is completed it may be desirable to visit the farm to complete the investigation, to investigate environmental aspects, although distance or expense may make such a visit impossible.

5   Beware of confusion if a new disease is superimposed on a pre-existing condition. In poultry work there are normally several carcases from a problem flock to counteract this hazard.

Techniques

A   Killing birds for autopsy: the following methods may be employed:
    1   Intravenous or intracardiac barbiturate.
    2   Dislocation of the neck.
    3   Carbon dioxide.

B   Autopsy
    1   Record weight, age, sex, breed and bodily condition.
    2   Record external signs or lesions, skin parasites, etc.
    3   Moisten feathers.
    4   Reflect the skin from the abdomen to the head and remove the sternum.
    5   Examine the respiratory, gastro-intestinal and genital tracts, the skeletal and nervous systems and other relevant organs.

6   Assess the findings in the light of the history and decide what further diagnostic tests are appropriate. Consult specialist laboratories on the selection of samples for further investigation.

7   Report to the farmer on preliminary findings and any treatment options and discuss the findings in relation to the developments on the farm.

E   Practitioners and Research Workers

Poultry practitioners have close, symbiotic relationships with research workers whom they often consult about emerging problems and supply with field data and samples for research. The research workers may be at the Central Veterinary Laboratory, a Veterinary Investigation Centre, a University, poultry or pharmaceutical industry laboratory, a government-financed research institute, or other institute.

Journal such as *Avian Diseases, Avian Pathology, Poultry Science,* and the *Veterinary Record* are perused by poultry practitioners seeking answers to their problems. Magazines such as *Poultry World* in Britain and *Poultry Digest* are useful sources of general information, and the former is authoritative as a source of marketing and political developments.

F   Poultry Farmers' Point of View

Disease hazards are a major source of worry for poultry farmers and they are interested in alleviating these worries by obtaining an early opinion on any problems that occur and prompt advice on what options exist for treatment.

G   Text books as an aid to diagnosis

Anticoccidial Information including safety, toxicity, incompatibilities and associated matters. Fowler, N. G., Canterbury, Antec Associates. 1995.
A Colour Atlas of Diseases and Disorders of the Domestic Fowl and Turkey, 2nd edition, Randall, C. J., London, Wolfe Medical Publications, 1991.

Compendium of Data Sheets for the Veterinary Profession 1994-1995, Anon, National Office of Animal Health, Enfield, EN2 7HF.

Diseases of Gamebirds and Wildfowl, Beer, J. V., Fordingbridge. The Game Conservancy, 1990.

Diseases of Poultry, 9th edition, eds. Calneck, B. W., Ames, Iowa State University Press, 1991.

Handbook of Medicinal Feed Additives 1994/5, 12th edition, ed. Mounsey, A. D., HGM Publications, Baslow, DE45 1RZ.

The Health of Poultry, ed. Pattison, M., Longman Scientific and Technical, London, 1993.

Isolation and Identification of Avian Pathogens, 2nd edition, eds. Hitchner, Domeruth, Purchase, and Williams, American Association of Avian Pathologists, Texas A & M University, 1984.

Poultry Diseases, 3rd edition, ed. Jordan, F. T. W., London, Bailliere Tindall, 1990.

Poultry Health and Management, 3rd edition, Sainsbury, D., Blackwell Scientific Publications, London, 1993.

The Veterinary Formulary: Handbook of Medicines Used in Veterinary Practice, 2nd edition, Debuf, Y. M., Pharmaceutical Press, London 1994.

Virus Diseases of Poultry, eds. McFerran, J. B. and McNulty, M. S., Elsevier Science Publishers B.V., Amsterdam, London, New York and Tokyo. 1993.

A vast range of textbooks concerned with many disciplines, from anatomy to zoology, may have to be consulted in the course of a year's work.

Types of Flock

A   Breeding flocks

Poultry breeding companies compete vigorously for large international markets for their products. Having selected the genetic material, grandparent flocks must be created free of major egg transmissible pathogens - *M. gallisepticum, M. synoviae, M. meleagridis, Salmonella pullorum, enteritidis* and *typhimurium* (and other salmonellae so far as possible), and leukosis virus. All breeding flocks are subject to salmonella monitoring under government supervision. Continuous monitoring by serology and culture is required, not least for export certification. Breeding companies require laboratory facilities of their own, with specialised help on techniques from research centres. They require high standards of isolation for the poultry units and staff. Disease 'breaks' do occur in practice but have to be contained.

Parent flocks derived from the breeding company will be in different ownership and may not adopt the same high standards of isolation. Such flocks will be in the Poultry Health Scheme (see below) as are virtually all breeding flocks and hatcheries.

Fertility: any evidence of 'infertility' will lead to investigations of embryos, eggs, diet, male birds, general flock health and artificial insemination (AI) techniques in the case of turkeys.

Egg hygiene involves frequent collection of clean eggs, fumigation, suitable storage conditions and incubator hygiene.

Egg washing (and egg pressure dipping in antibiotic solutions to eradicate mycoplasma) have been employed. Any egg washing system requires skilled staff to avoid contamination 'breakdowns' with these solutions.

The Poultry Health Scheme accepts breeding flocks, hatcheries and rearers which conform to specific health standards. It provides official recognition of the members flock health status thus facilitating trade and export certification. Where appropriate, screening for Salmonella gallinarum, pullorum, and arizona may be required also for Mycoplasma gallisepticm and meleagridis. A financial contribution from the members is now required for these services provided by the government.

The poultry industry likes the official 'blessing' this scheme appears to give to its operations and the scheme facilitates exports. Laboratory tests are available at Veterinary Investigation Centres at a special members' rate.

Exports: Veterinary practitioners issue health certificates in their role of Local Veterinary Inspector, subject to clinical investigations, laboratory tests and flock history, Protectionism can lead to excessive scruples re: disease hazards by importers.

Imports: Import regulations vary according to the country of origin. Since 1992/93 imports from the member states of the EU will be on licence only and this will not usually involve quarantine in UK. Imports from non-EU (Third) countries will also be on licence but will usually be required to spend 35 days in quarantine.

Cage birds, captive birds and pet birds present a risk of virulent Newcastle Disease and are therefore quarantined for 35 days.

The State Veterinary Service should be consulted on import (and export) regulations as new disease problems frequently require special conditions to be imposed, or in some cases the importation will be prohibited.

Migratory birds enter and leave the UK at will and may have introduced viral infections from time to time.

B Chicks and growers

Management skills and good housing are of great importance in giving young birds a good start in life. Each age range has its particular problems, several of which may occur simultaneously in the one flock.

Immunisation procedures have to be incorporated into the rearing period, in accord with the needs of the ultimate owner and the risks of the site for which they are destined.

Some of the more common disease problems that may occur are listed below:

1   Conception to hatching: dead in shell embryos: egg or incubation faults, nutritional defects, infectious agents.

2   Hatching to 10 days of age
    (a) Weak chicks: due to incubation defects, genetic defects, smallness; starve-outs are chicks which fail to feed and die within 4 days.

    (b) 'Chilled' chicks: transport difficulties, draughty lorries, air-strips etc. may stress chicks.

    (c) Environmental faults: inadequate or excessive heat, poor ventilation, atmosphere, etc., carbon monoxide fumes, human errors - may present as 'coli-bacillosis' or air sac infection.

    (d) Infected chicks

    (i) Environmental bacteria - umbilical infection, yolk sac infection (omphalitis) associated with *E. coli, staphylococci, Cl. histolyticum, pseudomonas* - may also be combined with salmonella infection.

    (ii) Pathogenic bacteria
    Salmonella pullorum is still present in some 'back-yard' flocks but is otherwise eradicated.

Salmonella enteritidis and *typhimurium* (regarded as invasive strains) are reduced by a government screening and slaughter policy for breeding flock. Other salmonellae are controlled by general hygiene methods on the farm and in the feed industry.

Other salmonellae, e.g. *S. hadar*, may be present with no disease or may cause septicaemic disease, especially if large dose of infection and susceptible stressed checks as in 2(a), (b) & (c) above. Such infections may come via the eggs or via the diet. Competitive inhibition concept is being employed: protection of the chick intestine from salmonellae can be achieved by adherence of certain non-pathogenic bacteria within a few hours of hatching. Antibiotic treatments may favour the carrier state for salmonellae.

(iii) Brooder pneumonia: aspergillosis - a sporadic problem.

(iv) Acute form of Runting Syndrome: deaths occur from 4 to 10 days.

3   <u>11 days to 11 weeks</u>
Classical forms of Runting Syndrome.
Nutritional deficiencies.
Mycotoxins and other toxins, or mistakes in mixing of feed (samples may be collected for analysis if necessary).
Coccidiosis and related problems.
Necrotic enteritis.
Marek's disease.
Infectious bursal disease.

Adverse vaccinal reactions may follow the administration of live vaccines by aerosol (or by water) if the environment is less than perfect. Viral infections, including particularly the immunosuppressive infectious bursal disease and Marek's disease. Bacterial and mycoplasmal disease may be facilitated by environmental defects - a common manifestation of which is the so-called 'coli-bacillosis'. Behavioural problems - cannibalism, and tendon injury resulting in lameness and incapacity.

4   12 weeks to point of lay
Marek's disease: despite vaccination breakdowns do occur and tumours may appear in muscle, ovary, nerves and elsewhere.

Leukosis - chickens over 16 weeks may be affected.

Footnote
In a survey of mortality in 9 flocks of replacement chickens (not broilers) totalling 60,403, 977 died (1.61%) between hatching and 70 days. Curtis, P E and Gabaj, M K (1986), *Proc. Soc. Vet. Epid. and Prev. Med.* Ed Thrusdfield, 174-182.
Major groups of disease were as follows:

Mortality

A   Umbilical infection/omphalitis   0.24%
B   Lung congestion, pericarditis, perihepatitis, coli-bacillosis 0.25%
C   Starve-outs, runting syndrome, runts 0-21 days   0.16%
D   Coccidiosis, peritonitis, necrotic enteritis, torsion of bowel,
      dehydration (post 15 days)   0.26%
E   Encephalomalacia (one flock only)   0.20%
F   Some 40 other conditions   0.50%

Disease and mortality were contained at an acceptable level by skilled work and regular monitoring of losses. The encephalomalacia affected only one flock killing 1.83% between 4 and 8 days despite adequate Vitamin E in the diet.

C   Broiler chickens
Broiler chickens are grown in vast numbers (some 600 million per year) in the UK and are a major source of meat. Most will be slaughtered by 7 weeks of age and although they are susceptible to the conditions found in young replacement chickens they have additional problems associated with their rapid growth potential, such as ascites and sudden death (or

flip-over) syndrome. Environmental factors are important in determining the mortality level and 6% is not unusual in the absence of major infections such as the more virulent forms of infectious bursal disease virus.

In a survey of two 20,000 bird broiler flocks (Curtis 1989) to 45 days the mortality, including culls, was 3.45% overall. The major diagnoses are shown below and are fairly typical of the usual problems, but it must be appreciated that each could assume greater proportions in other flocks.

Sudden death syndrome	0.94%
Yolk sac infection	0.39%
Ascites	0.34%
Runts	0.34%
Respiratory syndrome	0.42%
Leg problems	0.15%

Ref: Curtis, P E, (1989) *Proc 3rd European Symposium on Poultry Welfare, Tours.* Ed Faure, J M and Mills, A D, 249-251.

In 1995 broiler mortality in UK will regrettably be much higher as a result of the more virulent strains of infections such as Infectious Bursal Disease and Marek's disease which are circulating. Both these viruses are also capable of facilitating other disease conditions by immunosuppresson.

Skilled stockmanship is required to cope with the vagaries of buildings and climate and errors will quickly produce a disease episode. Litter management is essential to avoid hock and breast burns and foot lesions which are both a welfare and economic problem. Adequate insulation of buildings is vital. Ventilation may be by natural means with automatic control, or by fans, but neither method is foolproof although the former is gaining favour. The limitations of the broilers' respiratory efficiency requires the avoidance of pockets of still or stale air. An environmental medicine approach is essential to an understanding of broiler problems. Judgement will be required to assess what medication is necessary for treatment or prophylaxis and there is increasing interest in growing broilers with greater emphasis on hygiene and less reliance on medication, except when strategically necessary.

Reliance on medication to correct any errors is not advisable.

D Layers
Layers exist to produce eggs, that consumers will wish to purchase and convert into tasty nourishing high quality food.

I Production Problems
(a) Egg numbers, egg size, egg quality (shell, white, yolk, flavour) may be involved in production problems.

(b) Egg Drop
Any departure from the expected egg production chart involving a fall in production is called an 'egg drop' and producers have to seek advice immediately as to its cause, which could be any one of many factors - feeding, nutrition, husbandry, climate, egg stealing, bird quality or infectious disease. The farmer will probably consult several persons including his veterinarian who may undertake autopsies initially or serology.

II Disease Problems
Generally speaking, the UK laying flock has achieved a good health status with regard to most major infectious diseases, whether viral or bacterial. Turkey rhino-tracheitis virus may be having some effect, ILT is said to be widespread but of low virulence, and infectious bronchitis is affecting egg production and egg quality. The appearance of *Salmonella enteritidis*, phage type 4, in the UK in 1987 led to public alarm and resulted in a control scheme, which has partially restored public confidence in eggs. Egg hygiene has been greatly improved.

(a) Gynaecology

Large numbers of hens (1%+) inevitably succumb to oviduct-associated problems - these include impaction by whole egg, ruptured egg shell, or egg material, and facilitate ascending infection leading to peritonitis, and bacterial perihepatitis and pericarditis. Such infection may occur without apparent impaction but is regarded as of environmental origin. In other cases eggs are found intrra-abdominally placed. The right oviduct may persist as a small 'cyst' but can become distended and infected. Leiomyomas are quite common in the ovarian ligament.

(b) Other Diseases

Visceral gout: may reach epidemic level in some flocks - check on water supply and diet.

Accidental injury: fractures, crushing, etc.

Smothering: Fear and sudden shock (litter)
Crowding behaviour (litter)

Ammonia blindness - corneal ulceration - some days after exposure to high levels of ammonia following a cold spell externally.

Cage-layer fatigue: healthy layers become paralysed - Ca withdrawal from skeleton alleged.

Temporary partial paralysis affecting layers in breeding flock - injury by cockerels suspected - recovery in 24 hrs to 10 days. Marek's virus may be involved.

Vices   - cannibalism: can be severe problem especially if dietary changes or deficiencies - injured or paralysed birds may be pecked by their fellows.
     - feather pecking: feather loss is a major complex problem of modern hybrids - genetic susceptibility.

Parasitic conditions: red mite - monitor infestation level and spray cautiously. Roundworms - *Ascaridia galli* and *Capillaria* may occur but not in caged layers.

Many other problems occur from time to time - fatty liver haemorrhagic syndrome, adenocarcinoma of mesentery and other tumours.

Concern about drug residues in eggs inhibits medication.

E   Turkeys

Turkey rhino-tracheitis appeared in the UK in 1985 having been previously reported in Europe and Israel. The infection spread rapidly and caused considerable losses on some sites. The virus was identified and a vaccine is now available. The UK turkey industry has a good health status with regard to mycoplasmosis and to salmonellosis. Leg problems are commonly encountered.

Fowl cholera (*P. multocida*) is still a major hazard for the industry and some outbreaks can be very difficult if not impossible to control.

Erysipelas occurs in fattening turkeys as a clinical problem. Aortic rupture is a cause of sudden death in adults.

Viral conditions in addition to rhino-tracheitis include haemorrhagic enteritis, Newcastle disease, paramyxovirus 3, influenza virus including the virulent manifestation, fowl plague, for which there is a slaughter policy in force in the UK.

Chlamydial infections occur and parasitic diseases include coccidiosis and histomoniasis, both of which are usually controlled by 'in feed' drugs. Syngamiasis occurs in free range turkeys.

Turkey oedema disease occurs in turkeys, suspected to have a multi-factorial aetiology.

There is a view that with first class husbandry and management, turkey disease problems are easily surmountable, but human errors will always be a possibility, and new disease problems will continue to occur.

F   Ducks and geese

I   UK Ducks
(a) Specialised industry: e.g. Cherry Valley Farms Ltd., Rothwell, Lincs. breeds, feeds, cooks and exports ducks in every form to the world. There are other large duck companies.
(b) The rest: small flocks, pond and park flocks.

II   Infectious Diseases
Infectious diseases affecting ducks include the following:

Bacterial disease - anatipestifer disease, fowl cholera, *E. coli* infections, chlamydiosis, botulism, salmonellosis.

Viral conditions - duck virus enteritis, duck virus hepatitis, avian influenza, Newcastle disease.

Other conditions: lead poisoning, mycotoxins, coccidiosis.

III   Geese
The goose industry is not well developed in the UK. Geese are susceptible to many of the conditions affecting ducks. Derzsy's disease is a highly fatal disease of goslings caused by a parvovirus. Goslings are sometimes inadequately fed and die from starvation due to reliance on grazing without supplementation. A common parasite is the gizzard worm.

G   Backyard flocks, domestic and ornamental poultry

Introduction
In contrast to the sophisticated intensive poultry industry in Britain, there is another form of poultry-keeping which is often atavistic and

primitive and has little regard to the hygiene concepts of the commercial industry. No one knows how many small flocks exist but they appear to be on the increase, and are frequently kept for interest or hobby reasons.

Owner types
Historically in Britain chickens were kept for sport (cock fighting) or ornament rather than for food. The 12th Earl of Derby 1752-1834, had 2,-3,000 cocks put out 'to walk' annually and the whole population followed the 'sport'. Breeding show birds and exhibiting them attracts many diverse types of person who usually have little interest in, or knowledge of, disease. Ignorance is also quite widespread among genuine backyard poultry keepers, although with good husbandry, high levels of egg production can be achieved if sound stock is obtained.

An example would be a flock of 4 hybrid layers which in 1982/83, laid 1205 eggs in 365 days eating 225kg feed - 301 eggs per bird (82.5% production). Food consumption = 187g per egg.

Sources of stock
Stock is often obtained in a very casual manner from rearing agents, markets, acquaintances and even from pet shops. Hens from commercial flocks due for slaughter may be acquired by 'back yarders'.

Preservation
The Domestic Fowl Trust, near Stratford-on-Avon, was started in 1975 to conserve rare breeds of poultry.

Husbandry
Small flocks are frequently housed in primitive, improvised structures, with unsuitable or non-existent litter, dirty water supply and inappropriate food. Well-meaning but inexperienced owners can inflict considerable discomfort on the animals under their care.

Owners' questions
These are often concerned with the unsatisfactory performance of their domestic poultry. Why has laying not commenced - why so few eggs -

can they survive on scraps - do they need light - will modern hybrids thrive in my garden - is it cruel to keep them shut up? In addition to flock disease problems, individual pet birds are presented for diagnosis and treatment.

Disease problems

Small flocks in general are usually exposed to a wider range of infections than are the modern commercial units. See Curtis and Boachie (1982) 'Survey of the health and husbandry of small poultry flocks in Great Britain', *Veterinary Record* 117:216-219.

The results of this survey may be summarised as follows:

	Flocks
Proprietary food fed	12/15
Vaccinated	1/15
Losses due to foxes have occurred	5/15
Mycoplasma gallisepticum culture positive	8/15
Mycoplasma synoviae culture positive	7/15
Salmonella pullorum infected	1/15
Newcastle disease serologically positive (recent disease)	1/14 unvaccinated
Infectious bronchitis serologically positive	6/14 unvaccinated
Scaly leg (*Cnemidocoptes mutans*)	3/15
Body louse (*Menacanthus straminens*)	5/15
Fluff louse (*Gonicotes gallinae*)	1/15
Red mite (*Dermanyssus gallinae*)	1/15 (underestimate)
Adenovirus serology positive	9/15
Reovirus serology positive	8/15

Comment

Small flocks may be beautiful but they are also prone to acquire unusual infections.

Poultry markets
These provide an ideal opportunity for infections to be transmitted, especially as many would-be purchasers handle the birds to assess their quality. Poultry shows may also spread infectious disease, especially viral conditions.

Free range
Many small flocks are 'free range' but there are now an increasing number of commercial free range flocks being started to satisfy the consumer's desire for 'natural eggs'. Practitioners are reporting disease and welfare problems associated with this traditional system and fear that failure to rest the land will result in disease associated with parasitism and bacterial infections.

H   Game birds
Pheasant, partridge, and grouse are of considerable economic importance to the leisure industry, and there may be some 20 million pheasants in Britain at certain times. A mixture of traditional and intensive husbandry methods is employed to rear the birds, which are then introduced and acclimatised to the 'natural' habitat. Skilled gamekeepers are engaged to manage the operation on country estates.

Disease problems that occur in game birds are more likely to be the result of complex interactions involving several factors rather than a single factor, although one such may predominate. It must also be remembered that much less is known about game birds and their diseases than in the case of chickens. Environmental medicine is therefore the recommended approach for veterinarians asked to investigate a problem.

1   History
A full history should be obtained before any examinations are undertaken. Ensure that the species involved is correctly identified. It is essential to know the nature of the perceived problem and what level of investigation is required. In some cases a simple investigation is all that is called for, whereas in other cases a prolonged investigation involving

specialist laboratories may be required. Awareness of recent climatic changes in the locality is essential.

2   Autopsy

Dead or sickly birds may be presented as a first step in the diagnosis and they should be examined for evidence of injury or predator damage. Note particularly the body weight and condition, the amount and type of any food present in the alimentary tract, and if necessary check microscopically for gut protozoa with fresh (coverslip) preparations from intestinal or cloacal material. Cloacal material from live birds may reveal protozoal activity. Preliminary findings should be communicated to the client to consider what immediate action might be required to restrict losses, and to consider if the specimens received have been representative of the main problem, if there is one, and not simply part of the expected 'normal' mortality.

3   Environmental Medicine Consideration

Autopsy findings have to be interpreted in relation to the many environmental factors impinging on the birds in their habitat. A mixture of 'specific' infections or infestations may be detected at autopsy and these findings must then be interpreted in the context of the reality on the rearing site or shoot. Breeding and rearing game birds is probably more hazardous than commercial poultry because of the much less 'controlled' environment of the game birds and their genetic and behaviourial differences. Increased stocking density can have a negative effect on health in general.

4   Specific Diseases

(Identifying specific infections and conditions should not be an end in itself, but an opportunity to investigate what other factors contributed to the disease episode.)

Viral Infections: Game birds may occasionally succumb to poultry viral infections such as Newcastle disease. This disease is now rare in the UK, although the pigeon adapted form (pigeon paramyxovirus 1) recurs in epidemic form in pigeons in various localities. Runting syndrome may affect game birds with an acute form from circa 4 to 10 days of age, with some small birds being noted at or after three weeks of age. Marble spleen disease is another presumed viral infection which causes sporadic losses in adult pheasants and it is attributed to an adenovirus. A rotavirus has been associated with enteric lesions in the first week of life. Other viruses such as corona virus, ILT, or reovirus may be acquired from poultry. Louping-ill affects both grouse and sheep.

Bacterial Infections: Game birds are susceptible to the bacterial infections found in poultry and are more likely to be affected with 'old-fashioned' diseases such as tuberculosis and pseudotuberculosis (yersiniosis) than modern poultry. They may be exposed to salmonella infections from broody hens (*S. pullorum*) or from environmental contamination *(S. enteritidis, S. typhimurium,* etc.) Mycoplasma infections may affect intensively reared birds following vertical transmission from infected parent stock. Erysipelas and pasteurellosis may occur, and the usual environmental bacterial problems of checks (*E. coli, clostridia, staphylococci*) may be encountered if the environment is adverse. Young birds may then develop such conditions as yolk sac infection (omphalitis) and peritonitis, pericarditis, and air sac infection.

Fungal Conditions: Aspergillosis may be the result of hatchery or later environmental contamination, plus stress. Candidiasis affecting the crop may be part of a wider problem. Mycotoxins associated with mouldy feed can have an immunosuppressive effect and precipitate mixed infections.

Parasitic Conditions: The free-living nature of game birds, their exposure to wild birds, and possibly over-used ground, means that parasitism can be expected. Clinical parasitism may cause death, or just stunt growth and impair feather growth, thereby enhancing the birds' vulnerability to adverse weather.

A. ENDOPARASITES
Coccidiosis occurs in game birds with several different *Eimeria* species for each game species. Caecal and intestinal forms occur but the strains

may be of low pathogenicity. Anti-coccidials (e.g. clopidol) can be incorporated in feed during the rearing period, but immunity must be allowed to develop.

Histomoniasis (lesions in liver and caecum) occurs in pheasants and partridges - poultry and earthworm act as reservoirs; dimetridazole in the food may be used as a preventative. (EU approval uncertain).

Hexamita meleagridis: protozoan found in enteric conditions in the pheasant involving the caeca and intestine. Heavy losses can occur in poults circa 3 weeks of age. Recently released birds may be affected, and older birds become carriers. **Fresh autopsy material** is essential for reliable microscopic recognition of the motile parasite. Furaltadone or dimetridazole medication is used.

Trichomonas phasiani: protozoan having similar effects to Hexamita in pheasant and partridge, affecting the caeca and lower intestine. Medication as for hexamitiasis.

Syngamus trachea: syngamiasis or 'gapes' develops within 10 days of the ingestion of eggs or larvae from contaminated ground, earthworms, millipedes, or centipedes. Game birds, chickens, turkeys and wild birds are affected. Control is based on allowing low levels of infection to confer immunity and then treating the birds to prevent clinical gapes developing. Drugs used: mebendazole, febendazole, flubendazole.

Heterakis and *Capillaria* species occur - the former is a vehicle for Histomonas and the latter may cause weight loss.

Trichostrongylus tenuis: partridge and grouse are affected by this caecal parasite with a direct lifecycle involving encystment of larvae on heather. Stocking/population density influence are important in relation to nutrition and parasite burden. Grouse populations fluctuate periodically and sometimes catastrophically; research into these cycles continues.

B. ECTOPARASITES

Mites (e.g. red mite), lice, and fleas may be found on birds. Ticks can infect grouse (and sheep) with louping-ill virus.

Behaviourial Problems

The less domesticated nature of game birds means that they are more aggressive and 'wild' than chickens. Constant vigilance is required to avoid such behaviourial problems as crowding leading to asphyxiation, head or feather pecking, and cannibalism.

Cannibalism and pecking: multifactorial aetiology - environment and food supply should be checked. Keepers may fit 'C' shaped plastic 'bits' in the mouth of pheasants at 8 days or somewhat later, as a preventive, removing them prior to release. Trimming the upper beak is also used to deter pecking. Lighting intensity may contribute to cannibalism problems. 'Spectacles' are fitted to some adult breeders to prevent egg eating.

Birds which fail to find food can be a problem during the first four days of life and also following release. Fractures of long bones have been associated in young partridges with lack of exercise.

Nutrition and Disease

This may be a question of considering if the birds found the food provided or whether an adequate food supply was available. Contents of crop and gizzard at autopsy will be useful evidence. If adequate food was available, was it of the correct type for the species involved and did it contain any necessary additives? Vitamin supplements may be given separately in cases of unexplained poor growth. Mouldy food as a source of mycotoxins is an avoidable hazard and the immunosuppressive effects of mycotoxins can predispose birds to various infections. Well-grown pheasants may not always fly satisfactorily and there is debate as to the reason - is it genetic, nutritional, or due to husbandry methods?

Predators and Poisons
Losses due to these factors may have to be considered. Predator damage can be detected at autopsy and may have been overlooked by the client. If pesticides or agricultural chemicals are suspected the Ministry of Agriculture, Fisheries and Food may be consulted. Insecticides may adversely affect insect populations utilised by game birds and thus affect young birds.

Seasonal Aspects
Breeding stock is caught in the late winter and penned up on clean ground. Eggs are collected and stored, incubated (pheasants 24 and partridge 23 days), to produce successive weekly hatches of poults in the vital May/June period. The poults are intensively brooded initially and gradually introduced to additional space and open air, depending on the weather, so that they can be successfully transferred to release pens in the woods by 6 weeks of age. The close seasons in Britain for shooting are 2/2 - 30/9 (pheasant), 2/2 to 31/8 (partridge), 11/12 to 11/8 (grouse).

Further Information
The Game Conservancy, Fordingbridge, Hampshire SP6 1EF (telephone 014245 682381) publishes a range of booklets on game technology designed mainly for field workers. See also list of textbooks on page 11.

Vaccination and Medication
Notes by I. R. D. Cameron BVMS MRCVS

A Vaccination

A number of vaccines are licensed for use in poultry in Britain. Northern Ireland has special conditions. Licensed vaccines are currently available for the following:

Newcastle disease	ND
Infectious bronchitis	IB
Infectious bursal disease (Gumboro)	IBD
Infectious avian encephalomyelitis	IAE
Infectious laryngo-tracheitis	ILT
Marek's disease	MD
Egg drop syndrome, 1976	EDS
Turkey rhino-tracheitis	TRT
Avian paramyxovirus 3 (turkeys)	PMV3
Pigeon Pox	POX
Chicken Anaemia virus	CAV
Erysipelas	
Pasteurellosis	
Salmonella enteritidis	
Coccidiosis	

In general, vaccination induces a good immunity, providing the vaccines are correctly stored and administered and immunity not compromised. Recombinant viral vaccines will be a future development.

Viral vaccines may contain live virus of an attenuated or non-virulent strain, or killed virus incorporated with an adjuvant, whereas bacterial vaccines are usually killed, adjuvant vaccines.

Each vaccination programme is tailored to individual farms' requirements, and to be effective these vaccines must be used in

31

conjunction with, and in addition to, other methods of disease control. eg. site hygiene. When devising vaccination programmes for young birds, consideration must be given to the level of maternal immunity that is present. As a result, in most cases, the first dose of live vaccine will be given at about 3 weeks of age when maternally derived immunity will be waning. Data sheet recommendations for vaccine use and storage should be consulted.

The poultry industry requires mass administration techniques wherever these are possible, providing always that efficacy is achieved. Live vaccines are usually given by aerosol spray, or via the drinking water-low labour cost administration: when vaccinating in the water system consideration must be given to such points as water quality, its temperature and residual disinfectant. Some live vaccines are given by injection (MD), by eye drop (ILT), or by coarse spray. Killed (dead) adjuvant vaccines are given by injection - high labour cost administration, but if polyvalent vaccine is used (ND, IB, IBD, or EDS, in various combinations) the unit cost is reduced. Self injection with adjuvant vaccines can cause unpleasant lesions. Emergency vaccines can be prepared for special problem flocks, subject to government approval and licensing.

In-ovo vaccination at the hatchery is a technique which is gaining in commercial acceptance (MD and IBD) and will probably become an accepted technique in the coming years.

B   Medication
There are essentially three reasons for the medication of poultry:

1   Stress medication e.g. the use of soluble vitamin preparations following vaccination procedures or the movement of birds or other stress.

2   Preventive. In-feed anticoccidials for intensively reared birds and anti-blackhead (histomoniasis) drugs for turkeys are examples

of this type. In these cases access to unmedicated food should be avoided otherwise drug intake will obviously be reduced.

The inclusion of organic acids in feed (e.g. formic acid and propionic acid) has been used as a means of controlling salmonella contamination with reported good results both by use in feed compounds and in raw materials, and by providing resistance to recontamination.

In the future, with growing consumer pressure for additive free produce, the use of coccidial vaccines may become attractive. At present they are used in broiler breeder flocks particularly. Competitive exclusion products are proving a significant advance in the field of salmonella control.

3   Therapeutic medication. Once a diagnosis has been made, specific therapeutic medication may be necessary. This may take the form of vitamins, anti-protozoal drugs, antibacterials, or antibiotics.

Points to consider when thinking of medicating poultry flocks with antibiotics are:

a   Bacterial sensitivity/resistance.
b   Efficacy.
c   Dose rate.
d   Drug licensing legislation.
e   Onset and duration of medication.
f   Route of administration.
g   Drug stability and interaction with other drugs which may be already in use
h   Tissue residues.
i   Cost.

Dose rates, especially for water administration, are best calculated on a weight basis. Water and feed consumption can vary greatly due to such conditions as illness, ambient temperature, or management feed/water restriction programmes, and this can result in inaccurate dosage, if feed/water concentration alone is used.

Drugs may be administered by the following routes:

1 <u>Drinking water</u>: advantages: minimal labour, less stress, sick bird will still often take water, disadvantages: uneven consumption may result in uneven dosage.

2 <u>In feed</u>: advantages: no farm labour involved unless moving bagged feed is involved and no stress for the birds;
disadvantages: start of medication may be delayed, difficult to alter the medication, and sick birds have a reduced appetite.

3 <u>Injection</u>: advantages: the exact dose can be given to each bird and sick birds can be individually treated; disadvantages: labour intensive, stress may be associated with catching birds, and careless injections may cause damage.

Often the best methods of medicating poultry is to commence with water administration and continue later with in-feed medication, if necessary.

Care must be taken when medicating therapeutically to avoid any drug interactions and incompatibilities with drugs normally included in the feed, such as anticoccidials.

All feed medication with a Prescription Only Medicine (POM) must be authorised by a Veterinary Written Direction (VWD), accurately completed, which would clearly state the PML involved as well as the POM product to be added. The withdrawal period prior to slaughter for food, or to sale of eggs, must be clearly stated, (and Data Sheets should be consulted), to ensure that consumers are protected from the possible presence of drug residues.

For information on drug availability the following publications, which are regularly up-dated, may be consulted: Compendium of Data Sheets, The Veterinary Formulary, and the Handbook of Medicinal Feed Additives (please see booklist for details).

Poultry Welfare

A Veterinarians

In Britain veterinarians are deemed to know about animal welfare, especially for the purpose of administering the Codes of Practice. Government veterinary officers administer welfare activities, carry out spot checks on farms, investigate complaints, and advise the Minister on poultry welfare - a politically sensitive subject.

Poultry are normally too numerous to receive individual veterinary treatment and consequently a more holistic style is required with emphasis on environmental medicine as the appropriate approach to flock health and well-being. Whereas poultry health is a well researched field, 'well-being' with its implications for welfare is a minefield of uncertainty and doubt. Dilemmas must, however, be accepted as inevitable at the present time as the industry is caught between the conflicting ideologies of low cost food production and high welfare standards.

The practising veterinarian has a duty of confidentiality to clients, and this applies to any adverse information acquired in the course of the practice. Experienced practitioners are well able to give a balanced opinion on welfare issue to their clients and farm staff, and can indicate possible improvements in existing systems.

B Pressure Groups

Many pressure groups and organisations have an interest in poultry welfare and new groups are constantly arising. Organisations such as the Universities Federation for Animal Welfare, RSPCA, Compassion in World Farming, and others, promote greater awareness of the needs of animals, but there are at the other extreme, groups which appear to have been infiltrated by individuals intent on destruction and violence.

C   Government activities

British governments have for many years encouraged intensivism with a view to producing cheap food and cutting the labour force on the farm but the present government is now embarrassed by welfare criticism. Following the Brambell Committee Report (1965) on animal welfare, Codes of Practice for poultry were established (1968). The very independent Farm Animal Welfare Council (FAWC) advises the government on welfare.

FAWC periodically reviews various aspects of poultry welfare and publishes findings which are a useful contribution to the continuing debate and may help the government of the day by 'defusing' the situation by indicating its complexities. A 1994 investigation was concerned with the welfare of turkeys and previous reports were concerned with broilers and layer chickens.

D   European Union

Individual countries have to accept poultry imports from other EU countries and therefore cannot readily impose expensive welfare standards on their own flocks.

Welfare lobbyists are very active in the EU, reflecting the demand for progress from West Germany and Denmark among others. Southern countries are less concerned with these issues.

Special Marketing Terms: those adopted by the European Union for egg (labelling) systems are as follows:

	Maximum allowed
Free range eggs	1,000 b/hectare
Semi-intensive	4,000 b/hectare
Deep litter	7 b/m^2
Perchery (Barn Eggs)	25 b/m^2
Cages (minimum)	450 cm^2/b

In Switzerland, a non-EU country, the government has imposed high welfare standards for layers and broilers, and battery cages are to be prohibited. Mixed farms are favoured and the size of units has also been restricted to preserve the environment and encourage rural employment. The Swiss model will be studied as a possible prototype for future poultry systems elsewhere in Europe, when environmental and social accounting will be part of the policy making process. Sweden is to ban cages in the future.

E   Public opinion

There is much concern about poultry welfare in Europe and the northern countries, such as Sweden and West Germany, would welcome the abolition of cages in their present form. In the UK free range eggs are highly regarded and command a substantial premium over the cage-produced eggs, which however are not labelled as such. It is not clear to what extent the popularity of 'free range' systems in the UK is limited to a particular section of the population while others are more interested in having a cheaper product. Politicians are sensitive to criticisms of their inability to improve animal welfare standards unilaterally in the UK because of the free market in Europe.

F   Research and welfare

1   Disease prevention and control. Health is the first step to welfare. Alternative (non-cage) systems will usually involve greater risk of bacterial and parasitic infections and antibiotic treatment of the former will create drug residue problems for the veterinarian and his client. Cage systems have however been associated with bone fragility problems thought to be due to lack of exercise.

2   Injury avoidance. Human error in design of equipment or in husbandry may cause severe welfare problems. The collection and transport of spent hens and broilers needs improvement.

3   Breeding and selection. Selection for improved performance may have contributed to problems such as ascites and leg weakness in broilers.

4   Productivity. The idea that anything good for productivity is usually good for welfare is not acceptable because good production can be obtained in doubtful systems.

5   Ethology and behaviour.
(a) Ancestor investigation: just how does the Red Jungle Fowl behave - relate to modern chickens.

(b) Preference testing: workers have allowed chickens choice in accommodation and estimated the strength of the choice. Others have invited birds to 'vote with their beaks' by pecking a button to determine cage size.

(c) Negative thoughts of some behaviourists: many scientific behaviourists believe it is impossible to investigate the subjective world of animals and consequently say that to talk of consciousness and suffering is unscientific.

(d) Behaviours in litter: nest selection, perambulatory activity, feather loss, feather pecking, fear, stress etc. are subjects for active investigation.

(e) Physiology and biochemistry: hopes of measuring stress in chickens by a series of tests are attractive. Confusion about what is stress.

(f) Analogies with human experience: is it respectable? (Anthropomorphism - zoomorphism - solipsism).

(g) Philosophy and Religion. Animal welfare is a growth area in philosophy and there is considerable debate and discussion on the validity of the use of any species of animals, including poultry. The conflicting attitudes of various world religions are much discussed. Feelings often run high and mutual tolerance does not appear to be a feature of the debate. There is need to bring philosophers and veterinarians into contact to help clarify the issues.

(h) Psychology. More sophisticated investigations required into chicken psychology.

(i) Morality. Many views exist, often diverse and sometimes extreme.

(j) Smaller family run farms can often provide more "care" and more employment than very large units.

Background Information

A Species

The term poultry, from Old French 'pouletrie', nowadays is taken to include chickens, turkeys, ducks, and geese, although other species may be included for certain legal purposes.

B Some UK Statistics

The poultry industry provides meat and eggs for a population of 57.8 million people.

Poultry meat consumption has been rising in recent years (unlike other types of meat), and the rise is expected to continue, encouraged by the health lobby.

Egg consumption has been falling for some 10 years notably following the *Salmonella enteritidis* alarms of 1988/89. The improved hygiene standards then introduced, maintained consumption. Free range eggs have become increasingly popular over recent years and non-cage eggs currently account for an estimated 12% of sales and attract a premium price.

Poultry meat consumption (1993)	21.5 Kg per person per year
Egg consumption (1993)	172 eggs/person/year
Poultry populations (1993)	
Layers	33.2 million
Growing pullets	10.7 million
Breeding chickens	6.7 million
Broilers	73.00 million
Turkeys	7.8 million
Ducks and geese	2.30 million
Placings per year (1988)	
Broiler chicks	669 million

| Layer type chicks | 32.2 million |
| Turkey poults | 38.2 million |

Social Trends (1994), London, HMSO
Annual Abstract of Statistics (1994), London, HMSO

C   Industry Organisation

The poultry industry, which is dominated by a small number of large companies, has experienced considerable swings in profitability in recent years. The business appears to be highly competitive and very complex with integrated companies engaged in a whole range of operations - breeding, rearing, growing, slaughtering, portioning and cooking chickens - for home and overseas markets.

Public interest in less intensive systems has, however, given small producers an opportunity to achieve higher prices for their products.

There are close links between the poultry industry and the pharmaceutical industry since the former has a considerable need for vaccines and other products such as anti-coccidials, which are available to poultry farmers without veterinary involvement. There are also close links with the feedingstuffs industry and some companies have a foot hold in both camps.

The poultry industry is very innovative and very responsive to changing requirements of the market. Since poultry only live for a relatively short time compared with other species, the lessons of one flock are quickly applied to subsequent ones, and consequently, new ideas and techniques appear to come along at a more rapid rate than elsewhere.

The post-GATT free market situation may prove a serious threat to the future of the UK poultry industry.

| CONDITION | EGG DROP SYNDROME (EDS'76, or EDS) |
| Species | **Chicken** |

Clinical aspects   No apparent signs observed in chickens but egg shell quality and colour are reduced and egg production falls as a result of infection in susceptible laying flocks. This leads to a period of erratic egg production followed by recovery. Drops in egg production of up to 40% have been described but lesser figures of 10% to 15% are more typical.

Age range   All ages of layers susceptible.

Causal factors   1   EDS'76 virus - this adenovirus of duck origin was introduced into poultry flocks via a contaminated live vaccine in 1976.
2   Vaccination - the disease has been virtually eradicated from the commercial flocks through the use of killed vaccine but may be re-introduced into free-range flocks by wild bird contacts.
3   Vertical transmission.
4   Contaminated pond water with wild duck contact.

Pathology   Uterine oedema described in experimental infections.

Diagnosis   Serology: HI test, ELISA
Virus isolation

Immediate action   Investigate possible sources of infection.

VIRAL DISEASES

45

Long term action	Either obtain future stock from EDS-free breeders (the infection is vertically transmitted) or vaccinate prior to start of egg production.
Economic	This important disease caused loss of egg production for several years - vaccination introduced the disease and a new vaccine virtually eradicated it.
Welfare	Poultry welfare not affected to any detectable extent.

CONDITION	AVIAN INFLUENZA
Species	Theoretically all avian species are susceptible but in farming turkeys are the main target. Migratory birds transmit strains of the virus internationally.
Clinical aspects	Great variations from very mild depression, decreased egg production, respiratory and ocular involvement to sudden high mortality. The latter is called fowl plague and is a notifiable disease in Britain, but occurs only very infrequently.
Age range	All ages susceptible.
Causal factors	Type A Influenza virus: many strains, reservoirs of infection in wild birds.
Pathology	In an acute case haemorrhagic lesions are observed and necrotic foci may be present in spleen and liver - in mild cases no specific lesions can be described.
Diagnosis	Virus isolation is required in the initial stages of an outbreak. Serology
Immediate action	Fowl plague is a notifiable disease. Confirmation of the diagnosis requires laboratory assessment of the virulence of the virus isolated, by tests at Weybridge Laboratory (Ministry of Agriculture, Fisheries and Food). Flock slaughtered if positive for highly pathogenic Avian Influenza virus. In milder forms of the disease in turkeys, attempt to confine the infection to one site. No specific medication. Vaccines not used except in mild disease in USA.

Long term	Wild birds may introduce infection. Quarantine controls some domestic poultry imports and captive pet bird imports.
Economic	In the USA massive outbreaks of fowl plague in chickens in 1983/84 were controlled by expensive government financed eradication by slaughter. Outbreaks in the UK would prejudice poultry exports.
Welfare	In winter wild birds short of food enter turkey houses and may introduce infection to turkey flocks.

CONDITION	FOWL POX
Species	**Chicken, turkey**

Clinical aspects Skin forms affect the featherless part of the head - in the mouth there may be diphtheritic lesions. Mortality usually low depending on the virulence of the strain and intercurrent infection.

Age range Any age may be affected. (The disease is not common in the UK but occurs on multi-age sites in the USA.)

Causal factors Pox virus (avian type).

Pathology Hyperplasia of the epithelium leading to the formation of papules and vesicles which may coalesce - lesions eventually heal.

Diagnosis
1 Clinical signs may be diagnostic.
2 Electron microscopy.
3 Virus isolation.
4 Serology: GDPT

Immediate action Cull sick birds.

Long term Vaccination of susceptible poultry with either pigeon pox or fowl pox vaccine is possible prior to exposure, but eradication of infection from a site by depopulation at the end of lay and disinfection is the preferred objective.

Economic aspects Moderate importance.

Welfare Infected chickens or turkeys will suffer considerable discomfort and pain.

Footnote

1   Live pox vaccines may be contaminated with other viruses and are therefore less than ideal.

2   Pox virus infections exist in many specie of birds including pigeons and canaries

CONDITION	INFECTIOUS AVIAN ENCEPHALOMYELITIS (EPIDEMIC TREMOR)
Species	**Chickens mainly affected - turkeys and pheasants less susceptible.**
Clinical aspects	Chicks: Ataxia, depression and fine muscular tremors, progressive paralysis, death ensues. Adults: No clinical signs but a slight drop in egg production (5%) is possible with reduced hatchability.
Age range	'Tremor': chicks 1-6 weeks. Lowered egg production: adult layers.
Causal factors	1 Infectious avian encephalomyelitis virus. 2 Vaccine strains can cause disease in susceptible stock. 3 Vertical transmission causes disease in chicks. 4 Lateral spread.
Pathology	Lesions of encephalomyelitis and peri-vascular infiltration of brain and spinal cord.
Diagnosis	1 History, signs, and histopathology. 2 Virus isolation using susceptible embryos or fluorescent antibody tests, may be required to confirm diagnosis. 3 Serology: ELISA
Immediate action	Kill affected chicks - no treatment available.
Long term	Protection results from vaccination of young breeders between decay of maternal immunity and onset of lay, with live vaccine in drinking water.

Economic aspects   Purchaser may sue breeder if disease appears in chicks.

Welfare   A disease to be prevented by vaccination.

CONDITION	INFECTIOUS BRONCHITIS (IB)
Species	**Chickens and pheasants**

Clinical aspects

Classical: IB is a severe respiratory disease with a 2-day incubation period, not much mortality but egg production drops and shell quality and egg quality are impaired. Nephritis occurs with some strains, especially in colder environment.

Modern: Well vaccinated flocks in Europe may suffer from erratic egg production which occurs without any respiratory signs, and newer strains of IB are involved. Other new strains may cause classical disease.

Age range

All ages susceptible.

Causal factors

1 Numerous strains of infectious bronchitis virus (IBV) exist and others may "emerge".
2 Multi-age sites are at risk.
3 Poor environment, or mycoplasma, may contribute to the pathology.
4 Vaccines may not protect against "variant" strains.

Pathology

Respiratory tract lesions in trachea and air sacs occur in classical disease, and even in mild disease may contribute to the development of coli-bacillosis.

Loss of shell colour in the UK and loss of egg quality everywhere, are major pathological changes, also some loss of egg production. These changes may occur without any respiratory signs.

Diagnosis	1 Serology - rising titre to serum neutralising, haemagglutination inhibition or ELISA tests. 2 Virus isolation. 3 Clinical and epidemiological signs.
Immediate action	Medication of chicks may be helpful: no treatment of adults unless complications likely.
Long term	Check vaccination techniques. Research into new or genetically engineered vaccines. Flock isolation from possible "variant" strains of virus.
Economic aspects	Modern form is a serious cause of sub-optimal egg production and poor egg quality. Enhanced virulences could be a problem in the future.
Welfare	Today farmer's welfare suffers rather then that of the chicken from this disease (in the UK).
Footnote	Need for improved vaccines - research in progress.

CONDITION	INFECTIOUS BURSAL DISEASE (IBD) = GUMBORO DISEASE
Species	**Chickens - in other species e.g. turkeys, pheasants not usually a problem**

Clinical aspects

An important aspect of this infection is its immunosuppressive effect resulting from Bursal damage which may facilitate the appearance of other diseases. Since 1989 a more virulent (intermediate virulence) strain of the virus has been causing significant mortality in UK broiler flocks and more serious losses in susceptible replacement pullets. Classical IBD has a speedy onset with wet litter, a sudden 'spike' of deaths (1-70%) over a short period, and a quick recovery.

Age Range

1-16 weeks, associated with the persistence of the Bursa (but more usually 3-6 weeks).

Causal factors

IBD virus
The virus is highly resistant to normal disinfection and is site associated but is not egg transmitted.

Maternal immunity following adjuvant vaccination of parents controls mild strains, but interferes with the more virulent live vaccine available to protect against intermediate strain infection. Adjuvant vaccine given to chicks is effective but troublesome to administer and hazardous for operators.

Pathology

Bursa of Febricius inflamed and enlarged. Haemorrhages in thigh muscles. Dehydration in later stages. Impaired immunity to other disease

55

and poor response to other vaccines.

Diagnosis	History, signs and autopsy.
	Virus isolation.
	Serology: agar gel, ELISA, other tests (SNT).
	Immunosuppressive manifestations may go undetected.

Immediate action — No specific treatment.

Long term — Eradication of the more virulent strains from infected sites is feasible and desirable, since otherwise they can be expected to increase in virulence with time, as has happened in the past.

Welfare and economic importance — Affected birds may suffer a very unpleasant death. Economically the disease has a severe effect on UK poultry flocks particularly since 1989 and it has caused problems in the developing countries, such as Nigeria to which it was inadvertently exported in the past. Any types of IBD infections will impair flock productivity.

CONDITION	AVIAN LEUKOSIS (AL)
Species	**Chickens**

Clinical aspects Malignant conditions affecting the lymphoid system. Paralysis is not a feature. Symptoms otherwise are variable depending on which organ affected.

Age range 12 weeks and onwards.

Causal factors Leukosis/sarcoma retrovirus
Genetic susceptibility.

Pathology Lymphoid leukosis - a common manifestation is 'big liver disease' - other forms are erythroblastosis, mycoblastosis, nephromas, haemangiomas, etc.

Diagnosis Gross pathology, histopathology, history, virus isolation, serology. Gross pathology plus history is usually adequate for sporadic cases.

Immediate action Individual affected chickens should be killed. No vaccines are available.

Long term Eradication from breeding stocks in progress and some commercial hybrids are now AL free. Further research desirable.

Economic aspects Infected flocks may have suboptimal performance to a slight degree but mortality attributable to the disease is sporadic and unlikely to exceed 5% of any flock.

Welfare Eradication from breeding stock is desirable. No evidence of human susceptibility.

CONDITION	MAREK'S DISEASE (MD)
Species	**Chickens**

Clinical aspects　Paralysis of one or more legs or wings and/or increased mortality. Also immunosuppression effect may predispose to coccidiosis.

Age range　May appear at 3-4 weeks but usually after 8 weeks, with major losses from 12-25 weeks. A "floppy chicken" manifestation in broilers was recently recognised (1992/93) in UK.

Causal factors

1　Herpes virus (MDV).
2　Genetic susceptibility.
3　Inadequate vaccination.
4　New strains of high virulence are appearing in the USA and elsewhere.

Pathology　Lymphoid tumours of ovary, lung, heart, liver, kidney, spleen, skin, muscle, etc - enlargement of peripheral nerves, mesenteric, brachial and sciatic plexuses.

Diagnosis　History, symptoms of paralysis and lesions usually suffice. Paralysis is often unilateral, affecting leg or wing.

Immediate action　Destroy affected chickens and retain the rest of the flock.

Long term　Vaccination of day old chicks at hatchery and avoidance of infection during immediate post-vaccination period. Herpes virus of turkey (HVT) or attenuated virus (MDV) are used for live vaccination but combined vaccines are required to control new highly virulent strains.

Economic aspects	Formerly caused heavy losses and may yet do so again if more virulent strains appear. Eradication of the virus is not considered feasible as it is ubiquitous.
Welfare	The discovery of Marek's disease vaccine resulted in a great increase in the number of chickens surviving and therefore in better welfare. More virulent strains of virus are a new threat to welfare in UK.

CONDITION	INFECTIOUS LARYNGO-TRACHEITIS (ILT)
Species	**Chickens particularly susceptible**

Clinical aspects Very mild forms: conjunctivitis with a very few deaths (0.2%) from upper tracheal obstruction. Acute disease: gasping plus 10-40% mortality. "Back yard" flocks sporadic deaths with tracheal plugs.

Age range All ages susceptible but is most often found in adult commercial layers.

Causal factors Herpes virus - latency and carrier state.
Multi-age sites. "Back yard" flocks.
Live vaccine use may encourage persistence of infection.

Pathology Tracheal epithelium inflamed and may be desquamated resulting in plugs which occlude the larynx and cause asphyxiation. Haemorrhage in trachea in severe cases.

Diagnosis History, clinical signs, autopsy, electron microscopy, virus isolation. Serology DPT, SNT.

Immediate action Vaccination of other flocks on the site - only if the disease is other than the very mild form.

Long term Eradication from the site - otherwise routine vaccination of young stock with live vaccines (eye drop and other routes) is required.

Economic aspects This is a manageable disease.

Welfare Eradication from sites or regions is desirable.

CONDITION	NEWCASTLE DISEASE (ND)
Species	**All birds susceptible. Chickens and turkeys particularly affected.**

Clinical signs

These are very variable depending on strain of virus and the extent of vaccination in the flock. Acute and sub-acute strains cause respiratory signs, drops in egg production, soft shelled eggs, greenish loose faeces, and some torticollis or other nervous signs in individuals. Per-acute strains cause sudden death.

Age range

All ages susceptible.

Causal factors

Newcastle disease virus (NDV).
Pigeon paramyxovirus 1. (PMV1)
Imported cage birds, especially psittacines.
Adequacy of vaccination of flocks.

Diagnosis

Epidemiological features.
Virus isolation.
Serology: rising titres to HI test or ELISA if vaccinated.

Immediate action

No medication advocated in layers, but may be used in young chicks. Live vaccine can be used in the face of an outbreak, in nearby unvaccinated stocks.

NB
1 Slaughter policy exists in Northern Ireland.
2 Notifiable disease in rest of the UK with compulsory slaughter only for highly pathogenic strains.

Long term	1 Control imports of poultry and cage birds. 2 Only mild vaccine (Hitchener B1) and killed adjuvant vaccine allowed in the UK (except Northern Ireland). 3 Keep pigeons out of poultry houses and poultry feed.
Economic aspects	A disease of major importance and concern worldwide which causes continuous heavy economic loss overseas.
Welfare	Britain is virtually free of this disease but the use of mild live virus vaccine may cause some slight air sac infection in young birds, consequently broilers are often not vaccinated - but they are susceptible. Welfare dilemma.
Pigeons	Pigeon paramyxovirus (PMV1) reached UK in 1983. It caused, and still causes nervous symptoms, diarrhoea and death in affected wild and domestic pigeons. Pigeon contamination of poultry food led to outbreaks of Newcastle Disease in some laying flocks in the 1980's.

CONDITION	RUNTING SYNDROME (INFECTIOUS STUNTING SYNDROME (ISS), MALABSORPTION SYNDROME (USA)
Species	**Chickens - broilers particularly, but also guinea fowl and pheasants, and turkeys to a lesser extent.**

Clinical aspects

An acute form causes mortality in 4-10 day old chicks, the affected chicks are thin and gizzards contain litter, but no specific lesions. Over 3 weeks of age restricted growth and helicopter wings are observed in the active runts in the flock. Very few affected birds in many flocks - maternal immunity may determine the number of cases in a flock.

Age range

From 4 days.

Causal factors

1 A transmissible agent, presumed to be a virus, which has yet to be identified, impairs intestinal absorption.
2 Poor hygiene and overcrowding of chicks facilitates lateral spread.
3 Chicks from young breeder flocks are more often clinically affected.

Pathology

Pancreatic atrophy and fibrosis is described, possibly due to obstruction of the pancreatic duct, following 'viral' inflammation of the duct.

Diagnosis

Epidemiological diagnosis will be based on history, autopsy and clinical signs.

Immediate action

Cull small birds.

Long term   Improve house hygiene and husbandry. More research required: much research since 1976 has been directed at identifying the transmissible agent involved. Eventually it may be identified.

Economic aspects   Major importance worldwide.

Welfare   The small birds may not suffer particularly if they had access to feed and water, since they are usually more active than the 'normal' broiler.

CONDITION	TURKEY RHINO-TRACHEITIS (TRT)
Species	**A new disease of turkeys (in the UK since 1985) which may cause some lesser problems in chickens and pheasants.**

Clinical aspects   Turkeys develop a sudden respiratory disease with ocular and nasal discharges and distension of the infra-orbital sinus. The virus is immunosuppressive and secondary bacterial/fungal infections of the lungs occur if ventilation/husbandry defective. Temporary drop in egg production in laying turkeys.

In chickens temporary depression, reduced egg production and lung/air sac lesions reported from the field - swollen head in broilers may be associated with this infection.

Age range   Any age may be affected.

Causal factors   A pneumovirus has been identified as the agent and airborne spread is rapid. Environmental and other factors will affect the severity of any disease which may occur.

Pathology   No specific lesions - various respiratory lesions.

Diagnosis
1   Clinical signs and epidemiology of outbreak.
2   Virus isolation in tissue culture.
3   Serological tests ELISA, SNT.

Immediate action   Clinical judgement will determine if anti-bacterial medication required - in mild cases it is of doubtful value.

Long term   Live, and inactivated vaccines have been developed. This disease can be expected to persist in the UK and remain a threat to the poultry industry.

Economic aspects   Heavy losses occurred in the UK in 1985/6 and the disease will presumably continue to 'smoulder'.

Welfare   This infection will be a welfare problem for turkeys housed in less satisfactory conditions and which consequently develop secondary infections and fail to thrive.

Footnote   This viral condition must not be confused with a quite distinct form of rhino-tracheitis in turkeys which has been attributed to infection with Bordatella species.

Rotavirus

Rotavirus infections of chickens and turkeys are widespread and are thought to contribute to some outbreaks of diarrhoea in young chickens and turkeys, resulting in wet litter, which may lead to skin lesions in broilers. Young pheasants and partridge also at risk.

Reovirus

Although reovirus infection is widespread in healthy flocks, some strains are pathogenic and are believed to cause teno-synovitis (or viral arthritis) of broilers and pheasants affecting especially the digital flexor and tarsometatarsal extensor tendons, resulting in lameness and condemnations at slaughter in the case of broiler.

Chicken Anaemia Virus

This unclassified very small virus causes avian infectious anaemia with aplastic anaemia and lymphoid atrophy, usually appearing at circa 10 days of age and manifesting as 'blue wing'. Immunosuppression or lack of maternal immunity will allow the disease to appear, but in most commercial flocks the infection is subclinical. Young breeder flocks may acquire infection early in lay and produce vulnerable chicks for a few weeks without any clinical signs in the adults. CAV infected flock, may have reduced productivity and CAV may enhance other infections.

Adenovirus (other than Egg Drop Syndrome 1976)

These viruses are widespread in healthy flocks and there is evidence of latency. There are also certain disease conditions associated with adenovirus:
Quail Bronchitis.
Haemorrhagic Enteritis of Turkeys: widespread infection but little clinical disease.
Marble Spleen Disease of Pheasants.

OTHER VIRAL DISEASES

Inclusion Body Hepatitis: in young chickens 4-8 weeks of age is of complex aetiology in which adenovirus may play a part with CAV contributing to the disease.

Lympho-proliferative Disease of Turkeys

Retrovirus infection affecting only turkeys - characteristically splenomegaly is present but lesions also occur in other organs.

Reticulo-endotheliosis of Turkeys

Retrovirus infection of turkeys causing sporadic losses in older turkeys with tumours in the liver, spleen, gonad, bursa, kidney, etc. Chicken vaccines contaminated with this virus caused disease in chickens overseas.

Duck Virus Hepatitis

Acute disease of young ducklings in Britain for which a live attenuated vaccine is available.

Duck Virus Enteritis (Duck Plague)

A sporadic acute herpes infection of chick, geese and water fowl which may be introduced by migratory birds to local areas.

Derzsy's Disease

Causes diarrhoea and deaths in wild and domestic goslings and in Muscovey ducklings after 5 - 7 days incubation.

Other paramyxo-virus Infections

There are may strains of PMV infections, (other than Newcastle Disease) in birds and poultry and these include PMV3 of turkeys.

CONDITION	CLOSTRIDIAL DISEASES
Species	**All poultry susceptible to botulism and necrotic enteritis. Ulcerative enteritis in quail, chickens and turkeys.**
Clinical aspects	1 Botulism: flaccid paralysis, limberneck, death. 2 Ulcerative and necrotic enteritis: illness, diarrhoea, death.
Age range	Botulism may affect any age. Necrotic enteritis: 2-8 weeks usually. Ulcerative enteritis: 4-10 weeks.
Causal factors	Botulism: *Cl. botulinum* Necrotic enteritis: *Cl. perfrigens* Ulcerative enteritis: *Cl. colinum*
Pathology	Botulism: no gross lesions. Necrotic enteritis: necrotic enteritis. Ulcerative enteritis: typhilitis and ulcerative enteritis with necrotic foci in the liver.
Diagnosis	Botulism: demonstrate toxin by ELISA if available. Necrotic and ulcerative enteritis: microscopic examination of lesions plus culture if necessary.
Immediate action	No treatment for botulism. Penicillin therapy rapidly stops necrotic enteritis. Ulcerative enteritis: check predisposing causes in the environment.
Long term	Improve hygiene and husbandry.

BACTERIAL DISEASES

Welfare   Enteric lesions lead to chronic peritonitis in some cases.

Footnote   Botulism in broilers is sporadic and may be unrecognised. It is usually due to Type C organisms, the toxin of which does not appear to affect human beings. Litter containing broiler carcases has, however, been implicated in outbreaks of botulism in cattle grazing fields to which the litter has been applied. The carcases should of course have been previously collected and destroyed.

CONDITION	FOWL CHOLERA (PASTEURELLOSIS)
Species	**Turkeys, chickens, ducks, geese - also other avian species**

Clinical aspects

Acute form - sudden death, live birds sick, with mucus discharge from mouth and nostrils.
Chronic form - wattle cholera (swollen wattles) in chickens, joint infections.

Age range

Traditionally in older turkeys (over 8 weeks of age) and chickens (over 12 weeks) but may appear in broilers, occasionally, from 3 weeks of age.

Causal factors

1 *P. multocida* infection.
2 Rat/bird interactions due to overcrowding or disturbances in flock - rat bite transmits infection.
3 Virulence of the strain of *P. multocida*.
4 Heavy breeds of chicken more often affected initially.
5 Infection can persist in a recovered flock - carrier state.
6 Caged chickens not affected by this disease.

Pathology

Essentially a septicaemic condition, the invasiveness and virulence of the organism determines the lesions, which are very variable.

Diagnosis

Culture is essential and straightforward. In peracute cases blood/organ smears may reveal characteristic gram negative organisms.

Immediate action

Isolate sick birds, destroy carcases and medicate water or food. Turkeys may need parenteral medication in the early stages.

Long term

Aim to depopulate the site eventually - eradicate rat populations. Killed vaccines are available but do not guarantee freedom from the disease. Emergency vaccines can be prepared for a farm site with a continuing problem.

Economic aspects

This is a major disease of the turkey industry. Problem sites may be forced out of production by recurrent infection. Total slaughter essential with high virulence strains.

Welfare

Chronic forms of the disease cause considerable discomfort to survivors.

| CONDITION | E. COLI INFECTIONS |
| Species | **All poultry** |

Clinical signs   Very variable, depending on organs affected.

Age range   Post hatching - any age.

Causal factors   Disease episodes are often the result of young birds being subjected to sudden climatic change, sometimes aggravated by poor insulation. Respiratory viral infections, including attenuated live vaccines, will also predispose to infection. Vent injury and egg or shell impaction predispose to ascending infection of the oviduct, and peritonitis in adult females.

Pathology

1   Some forms of yolk sac infection.
2   Coli-bacillosis, in which condition pericarditis, perihepatitis, air sac infection and pneumonia may be present.
3   Peritonitis in layers may be associated with 'ascending' oviduct infection. Salpingitis in ducklings.

Diagnosis   Culture plus pathology plus history.

Immediate action   Medication may be considered depending on flock picture and the sensitivity of the strain. Check farm environment and ventilation controls.

Long term   Check environment. Inter-flock hygiene on farm (disinfection) is vital.

Economic aspect   Serious cause of broiler deaths and condemnations.

Welfare   These conditions have caused a considerable
amount of chronic ill health leading to death in
chickens, both broilers and layer replacements,
and in other species of poultry.

CONDITION **Species**	AVIAN TUBERCULOSIS **All poultry susceptible**
Age range	Older birds more affected than young ones. Uncommon in modern. intensive flocks.
Causal factors	1 *Mycobacterium avium.* 2 Poor hygiene - exposure to wild birds. 3 Infected sites perpetuate infection: caged birds, less risk of infection.
Pathology	Tuberculous lesions appear in liver and spleen - bone marrow and intestines also often affected.
Diagnosis	Autopsy plus microscopic examination of ZN stained smears from lesions. Intradermal tuberculin testing is possible. Distinguish true tuberculosis from pseudotuberculosis (yersiniosis).
Immediate action	Remove affected birds and improve hygiene. Consider slaughter of the infected flock with eradication in view.
Long term	Buy in non-infected stock and attempt to prevent infection being introduced. This is difficult if on free range, since wild birds are heavily infected.
Economic importance	Of slight importance in the UK at present.
Welfare	Eradication is desirable, otherwise the disease persists on the site.

CONDITION	SALMONELLOSIS (OTHER THAN PULLORUM DISEASE AND FOWL TYPHOID)
Species	**Chicken, turkey, ducks, and other poultry and birds**

Clinical aspects Young chicks and poults may show clinical signs of septicaemia with yolk sac infection, foci in liver, and pericarditis. Deaths or poor growth may be noted. With less invasive strains, unapparent infection is a possibility.

Age range 0-6 weeks for clinical disease. Any age for carriers.

Causal factors
1 Infection with invasive strains such as *S. enteritidis* or *S. typhimurium*, or with other 'paratyphoid' strains.
2 Environmental stress may contribute to the disease.
3 Food contamination by rodents or via ingredients.
4 Vertical transmission via egg (shell or contents) from breeding stock.
5 Competitive inhibition: normal gut bacteria can inhibit the attachment of incoming salmonella.

Pathology Septicaemic disease may result but a 'healthy carrier' state is also possible.

Diagnosis Culture of lesions or intestinal contents is advised if salmonellosis is seriously suspected.

Immediate action All isolations of salmonella from livestock must be reported to the State Veterinary Service. Elite or grandparent flocks will be slaughtered if

S.enteritidis or S.typhimurium is involved. In other cases the clinician will have to consider the various options for the flock - treatments, nursing and hygiene, or slaughter. Medication may prolong the carrier state in some cases. Each case must be judged on its merits.

Prevention

Placing healthy stock in clean buildings, avoiding contact with carriers (including humans) and other sources of environmental contamination, can be reinforced by organic acid treated diets, and the use of 'competitive inhibition' additives.

Salmonella control the UK

Following a television programme shown on **in** 15.11.88 in which possible hazards of human infection with egg-borne *Salmonella enteritidis* was highlighted, a period of intense political alarm seized the population and embarrassed the government. (The more invasive forms of *S. enteritidis* had penetrated the poultry industry in both Europe and the USA in 1987.) The result was the resignation of one minister and a scheme for the compulsory swabbing of laying flocks, breeding flocks, and hatcheries, and the registration of these enterprises and their inspection by local or central government officials, with provision for slaughter and limited compensation if *S. enteritidis* or *S. typhimurium* was detected. Egg sales which had collapsed in 1989 slowly recovered as public fear was allayed by the scheme.

The original policy (1989) of slaughtering all infected flocks was replaced in 1993 by a scheme concentrating on breeding flocks of over 250 birds. It had come to be recognised that proper handling and storage of eggs was a major factor in limiting human exposure. Now only breeding flocks and hatcheries are subject to regular compulsory bacteriological screening. If Salmonella enteritidis or typhimurium is diagnosed the breeding flock is slaughtered, but parent breeders may be offered a treatment option. There is a voluntary Code of Practice for Salmonella control in poultry and there is also now an effective inactivated Salmonella enteriditis vaccine. Serological surveillance techniques are being evaluated.

Broiler growing flocks are not involved but infection in these flocks should recede in future as the broiler breeding flocks are subject to the scheme.

Control schemes for *S. enteritidis* are being developed in Europe and the USA.

Long term

Eradication of *S. enteritidis* is feasible but to eradicate other paratyphoid type salmonella from poultry may be too expensive to be cost effective. Irradiation techniques will also provide choice for those consumers wishing to buy salmonella free chicken meat.

Economic aspects

Salmonellosis is an important cause of loss of chicks/poults in less developed countries. Human food poisoning in Europe and the USA is a major problem to which all species of livestock contribute. Poor kitchen hygiene is a major contributory factor.

Footnote

Salmonella arizona is a major turkey pathogen in

the USA but is not present in British breeding
stock.

CONDITION	PULLORUM DISEASE AND FOWL TYPHOID
Species	**Chicken - less common in turkey - other avian species may acquire infection.**

Clinical aspects

Pullorum disease: sudden death or delayed deaths with signs of sickness, diarrhoea and depression, thirst and eventual death.
Fowl typhoid: sudden death or signs of sickness, diarrhoea and depression, thirst and eventual death.

Age range

Pullorum disease: young birds 0-3 weeks usually.
Fowl typhoid: older birds, 3 months onwards.

Causal factors

1 *Salmonella pullorum* and *Salmonella gallinarum*.
2 Poor hygiene and poor environment will cause enhanced losses.
3 Wild birds are a potential reservoir of infection, and increase in non-cage systems in the UK may involve risk of re-appearance of fowl typhoid.
4 Vaccination not permitted in the UK - used overseas but of limited value.
5 Small 'hobby' flocks of fancy strains are a reservoir of infection in the UK.

Pathology

Septicaemic lesions in acute cases - not pathognomonic.

Diagnosis

The Rapid Whole Blood Plate test is of value in screening suspect birds in flocks, at markets or sales. Culture from lesions is required to confirm the initial diagnosis.

Immediate action	Commercial breeding stock should be free of infection in the UK. Infected flocks should be voluntarily slaughtered but owner may not agree if show birds involved.
Long term	Poultry Health Scheme requires all flocks in the scheme to be free of infection but regrettably these diseases are not notifiable and eradication by government is not in prospect yet. Health scheme for small breeding flocks of pure breeds is lacking in the UK.
Economic aspects	These are two major diseases of poultry which have been eliminated from European and USA commercial flocks.

CONDITION	MYCOPLASMOSIS
Species	**Chickens and turkeys particularly at risk, but many avian species susceptible if intensively housed.**

Clinical aspects

These are variable and depend on virulence of strains and environment. Sinusitis, respiratory disease, conjunctivitis, synovitis and lameness may be seen.

Age range

Chicks and poults may be affected as embryos, as young birds and as adults.

Causal factors

1 *Mycoplasma gallisepticum*: chickens and turkeys.
2 *Mycoplasma synoviae*: chickens and turkeys.
3 *Mycoplasma meleagridis*: turkey.
4 *Mycoplasma iowae*: turkey (infertility).
   1, 2 and 3 have been largely eradicated from breeding flocks but 2 is now being found to be present on many multi-age commercial sites causing little evident disease in adults.

Inevitably environment and nutrition contribute to any disease which occurs, especially in turkeys with 3 when leg problems occur in rapidly growing turkeys.
Research into 4 continues.

Pathology

Lesions in infraorbital sinus, respiratory tract, air sacs, lungs, sternal bursa, joints and tendon sheaths, embryos (also impaired growth).

Diagnosis and screening	Breeding companies undertake regular screening using plate tests and other tests for serological screening to detect any re-infection which may have occurred. Cultural identification required since non-specific reactions occur with plate tests. The Poultry Health Scheme has standards for (1) and (2).
Immediate action	Treatment with antibiotics may be required. Infected breeding stock (with *M. gallisepticum*) would be slaughtered - the other mycoplasma (2, 3 and 4) are less important.
Economic importance	Mycoplasmosis is a major cause of economic loss through unthriftiness.
Welfare	Unthriftiness and lameness will cause discomfort.

Staphylococcosis	Staphylococcus infections occur in chickens as localised infections of tendon sheaths in the leg, often following sprain or injury, in foot or joint infections, and as a severe dermatitis associated with skin damage. Sporadic cases of septicaemia occur. Similar conditions affect turkeys.
Streptococcosis	Streptococci do not appear to be significant pathogens in UK poultry at the present time although individual birds may develop a septicaemia, and endocarditis occurs in broilers and in adults.
Infectious Coryza (Haemophilus paragallinarum)	This is a major disease of chickens which is not now present in the UK but may reappear with the increase in free range flocks. Conjunctivitis, nasal discharge, wattle swelling are the main symptoms. Do not confuse *Haemophilus parainfluenza* infection which appears when ILT is present, with true infectious coryza.
Chlamydiosis	*Chlamydia psittaci* is commonly found in ducks, occasionally in turkeys, and virtually never in chickens. Wild birds, psittacines and pigeons may be naturally infected. Outbreaks of disease/infection in ducks and turkeys have led to human infections.
	Tetracycline therapy may be indicated in disease in poultry and is essential in psittacines.
	Avian strains differ among themselves and are distinct from cat or sheep strains.
	Laboratory diagnosis:

OTHER BACTERIAL DISEASES

1 Ziehl-Neelsen stained smears.
2 ELISA kit.
3 Tissue Culture.

Erysipelas

Erysipelothrix insidiosa septicaemia should be suspected in cases of sudden death in fattening turkeys. Stained smears of heart blood may reveal gram positive bacilli - otherwise culture requires time. Treatment with parenteral penicillin is effective. Killed vaccines are available. The disease can affect other avian species - uncommon in chickens. Rat faeces are a source of infection.

Campylobacter Species

A condition known as vibrionic hepatitis exists in the USA but does not appear to be seen to any extent in the UK. Clearly many chickens (and other animals) are carriers of campylobacter species and these organisms may invade the avian liver occasionally, possibly as secondary infections, but in the UK they do not appear to be primary agents. They have been associated with loose droppings (wet litter) in young chickens, and can cause food poisoning in man.

Listeriosis

Although birds can be infected with *Listeria monocytogenes*, an environmental organism, the condition does not appear to be a problem in the poultry industry in Britain. Incorrectly stored cook-chill products containing poultry meat have been found to be contaminated, and the ability of the organism to grow at low temperatures has contributed to the problem, which was highlighted in 1989, following concern at *Listeria* contamination of cheese and salads.

Spirochaetosis	This tick-borne septicaemic condition occurs in warm countries but not in Britain. Chickens and turkeys affected.
Anatipestifer Disease	An acute 'coli-bacillosis' like condition of young ducklings which responds well to drug therapy. The causal bacterium is called (incorrectly) *Pasteurella anatipestifer* until its correct taxonomy is agreed. Diagnosis based on epidemiology plus bacterial culture.
Streptobacillus moniliformis	Found in arthritic turkeys associated with rat bites. Not common.
Yersiniosis (or pseudotuberculosis)	*Yersinia pseudotuberculosis* infection causes multiple yellowish caseous lesions in the liver, spleen and in other organs.

Affected birds are dejected, thin and infective for others via diarrhoeic droppings. Many wild birds, cage birds and pigeons are infected but chickens and turkeys are not commonly affected, unless kept under conditions of infection plus poor environment. Children may develop lymphadenitis, mimicking appendicitis, and zoo primates are very susceptible to the infection. Control is based on hygiene, for antibiotic treatment of infected poultry flocks is not considered to be of great value.

CONDITION	COCCIDIOSIS
Species	**Chickens and turkeys - also other poultry and game birds**

Clinical signs

Death, dullness, diarrhoea and possibly blood in droppings.
Also subacute forms - poor growth rate.

Age range

21-42 days is the common age range but older birds can be affected if no immunity has developed.

Causal factors

Oocysts are highly resistant and exist worldwide. Given moisture and warmth they will sporulate and become infective. Oocyst dose, environment and stocking density, will influence the significance of any lesions produced.

Control

Young chickens are usually fed anticoccidial drugs during the early rearing period. Immunity is induced by gradually reducing dosage, except in the case of broilers for which constant medication is usual. Alternatives to drugs are coccidial vaccines for chicks, and high level hygiene techniques, reinforced by strategic medication if required.

Pathology

Enteric lesions of varying severity depending on species of coccidia and other factors.

Diagnosis

Macroscopic lesions plus microscopic examination of gut scrapings for merozoites, schizonts and oocysts is sufficient. Necrotic enteritis (*Cl. perfringems*) and ulcerative enteritis (*Cl. colinium*) must be distinguished from coccidiosis.

PARASITIC DISEASES

Immediate action   Medication should be considered unless only a few individual birds are affected and the rest of the flock looks well.

Long term   The shuttle use of anti-coccidials on a farm discourages the development of resistance, in that different drugs are used from time to time with the same flock, or with different flocks. Coccidial vaccine are available and may be favoured by producers particularly if seeking the 'additive free' label for their products.

Economic aspect   A major disease of poultry which is still not fully understood epidemiologically. New drugs required to counter future resistance problems.

Welfare   Recovered birds may have chronic gut lesions and develop peritonitis.

CONDITION	MITES: RED MITE AND NORTHERN FOWL MITE
Species	**All birds**

Clinical signs Examine birds for northern fowl mite and other ectoparasites, but do not expect to find red mite on a live bird during daylight.

Age range All ages.

Causal factors
1 *Dermanyssus gallinae*: red mite.
2 *Ornythonyssus sylviarum*: northern fowl mite.
3 Failure of farmer to detect mites: explosive outbreaks occur.
4 Failure to treat infestations, possibly due to uncertainty re: drug residues in some cases.

Pathology Deaths occur in chickens from anaemia following heavy red mite infestation.
Regular blood loss may adversely affect egg production.

Diagnosis Red mite must be sought in the building, in cracks, under dried faeces, on perches, or a mite trap or detector may be constructed and placed in perching area. Regular checks required to detect unapparent infestation.

Immediate action Treat house when populations of mites increase - do not spray food or eggs with insecticide.

Long term Regular inspection to assess mite populations is required. Eradication is difficult since mites can live six months without 'food'.

Economics An undramatic drain on the chickens' resources.

Welfare   Mites cause discomfort to chickens and can also
                               attack humans in human habitations. Wild birds'
                               nests in houses or institutions lead to human
                               infestation. Explosive outbreaks of mite
                               infestations occur in the early stages of infestation
                               in poultry or people.

Cnemidocoptes mutans	A mite responsible for scalyleg, the very common infestation of backyard chickens.
Lice and fleas	Several species of lice may be found on backyard chickens but they are of low pathogenicity. Lice are also found on turkeys, ducks and geese occasionally. Fleas may infest chicken flocks.
Roundworms	1 *Ascaridia galli* is common in chickens not kept in cages but is not a significant pathogen. 2 Capillaria species are pathogenic and may cause unthriftiness in chickens in non-cage systems and in other poultry. 3 *Syngamus trachea* is uncommon in chickens, but attacks game birds and turkeys. 4 *Heterakis gallinarum* transmits Histomonas but is otherwise non-pathogenic. 5 *Amidostomum anseris* is the gizzard worm of ducks and geese and is pathogenic.
Hexamitiasis and Trichomoniasis	These protozoans have been incriminated in enteric conditions in game birds, and turkeys are susceptible to heximitiasis. Examination of fresh smears for motile protozoans is recommended.
Histomonas meleagridis	Causes 'blackhead' - caecal cores of necrotic material and characteristic liver lesions - a problem for turkeys which is controlled by in-feed medication or good hygiene. Chickens are rarely affected, other species occasionally.
Leucocytozoon Species	Not a problem in Britain but occurs overseas where suitable insect vectors (simuliid flies and midges) are present. Diagnosis by blood film examination.

Cryptosporidiosis This protozoan has sometimes been associated with respiratory disease in chickens, turkeys, other poultry and gamebird rather than the enteric disease associated with most avian coccidia. Avian cryptosporidia are not usually considered zoonotic.

CONDITION	ASPERGILLOSIS
Species	**All species of birds - chickens and turkey poults**
Clinical signs	Respiratory difficulties, gasping.
Age range	From hatching to 4 weeks of age is high risk period with 5 days of age a common onset time.
Causal factors	*Aspergillus fumigatus* plus immunosuppression or poor environment - or high dose factor as in brooder pneumonia.
Pathology	Characteristic lesions - circular white/green lesions on air sacs or respiratory tract. Lung pale foci 1-2 mm diameter.
Diagnosis	Demonstrate the mycelia microscopically but gross appearance of lesions is a very reliable guide in some cases.
Immediate action	Kill off the affected chicks or poults.
Long term	Check incubators and farm environment.
Economic aspect	A constant hazard, but less important in the UK now than in the past.
Welfare	Unthrifty chicks or poults.
Footnote	Wild bird hospitals and birds in zoological collections frequently acquire this condition. Human infections may occur in susceptible individuals.

FUNGAL DISEASES

Candidiasis

Is associated with 'thrush' affecting the crop, but the underlying cause may be defects in husbandry.

Mycotoxins

These may be present in the original feed as in the classical aflatoxicosis outbreak of Turkey X disease in 1961, or toxins may develop in mouldy feed encrusting storage silos or bins on the farm. Mycotoxins have been incriminated in problems of poor productivity, loss of appetite, and immunosuppression, etc., and should always be considered when investigating obscure syndromes.

CONDITION	BROILER ASCITES
Species	**Broiler chickens**

Clinical aspects Comb usually cyanotic, feathers ruffled, bird prefers standing, walks with difficulty, abdomen distended with fluid, sudden death. Losses range from less than 1% to 50% ore more, depending on environmental factors and age.

Age range 2 weeks plus, but individual cases in younger birds.

Causal factors Right ventricular failure due to circulatory problems associated with the inability of the modern avian lung to cope adequately with the oxygen requirements of rapidly growing broilers, particularly when environmental conditions place extra stress on the system. Defective air flow at bird level, excess carbon dioxide, cold spells, excess heat, dietary factors, all can precipitate an increase in the occurrence of ascites in the subsequent week or weeks. (High altitude farms are at risk overseas.) 60 to 70% of affected birds are males. Selection for rapid growth has created a problem related to respiratory function and genetic selection will presumably solve the problem given time.

A similar condition occurs in turkeys but at a lower frequency.

Pathology Some birds die before ascites develops fully, but otherwise the abdomen contains yellow fluid, with clots of fibrin and there is right ventricular dilatation, and hypertrophy of the right ventricular wall.

OTHER DISEASES

Diagnosis	Pathology, history, epidemiology.
Immediate action	Slaughter affected birds.
Long term	Selective breeding for increased respiratory capacity is essential. Improved ventilation at bird level dietary moderation and the capacity and ability to adjust to climatic changes without prejudicing ventilation are key factors.
Economic importance	It is possible that 1% of the national broiler flock is lost with this condition each year. Individual flocks may suffer much higher losses, particularly small flocks in winter, - and valuable well grown birds close to slaughter weight may be lost.
Welfare	Affected chickens suffer considerable discomfort but are often found on small farms with a good welfare record.

CONDITION	NUTRITIONAL DEFICIENCY OR DEFECT
Species	**All poultry suffer from these deficiencies at some time or to some extent; minor deficiencies are more likely than the gross deficiencies described in text books.**

Clinical aspects

Not possible to be precise about clinical manifestations in the real world because the situation is often complicated by environmental or infectious conditions.

Age range

All ages susceptible.

Causal factors

1 Diet mis-planned.
2 Poor quality.
3 Mistakes in mixing still occur.
4 Ingredients accidentally omitted.
5 Wrong food for the species.
6 Sudden changes in diet composition.

Food manufacturers in the UK are not obliged to describe actual ingredients of feed.

Pathology

Abnormalities of behaviour or function may become apparent and will lead to autopsy work and histopathology as in encephalomalacia due to Vitamin E deficiency, or rickets due to Vitamin D deficiency.

Diagnosis

It is customary for farmers to retain a sample of each batch of food delivered to a farm for subsequent analysis should food be under suspicion. Clinical and pathological findings may indicate a possibly specific dietary defect which would then be investigated in the food sample.

Immediate action	Change diet.
Long term	Discuss with food supplier.
Economic aspects	Very important especially in developing countries where erratic food quality is a major problem.
Welfare	An adequate diet is a major welfare objective.

CONDITION	BEHAVIOURAL DISEASE
Species	**All species**

Clinical aspects Typical examples of this group of problems are:
1   Feather pecking.
2   Cannibalism.
3   Bullying.
4   Aggregation behaviour causing sudden death.
5   Cockerel fights and chicken disputes leading to injury, including tendon injury and lameness.

Age range All ages.

Causal factors Social deprivation, stocking density, nutritional defects or sudden changes in diet, environmental problems, excess lighting, innate behaviour problems, or interactions of one or more of these factors, must be considered when investigating problem flocks.

Pathology Depends on the condition.

Diagnosis This will depend on assessment of the disease and its pathology in relation to the behaviour of the bird. Observations of flock behaviour are time consuming, and appropriate, inconspicuous clothing is desirable.

Immediate and long term action More research required into animal behaviour. Skilled stock keepers pre-empt problems.

Economic importance Feather loss is an important cause of increased energy expense to the farmer and cannibalism can cause significant mortality.

Welfare   Some of these conditions may be the result of
normal behaviour in an imperfect intensive
environment. Beak trimming is used in some
chicken and turkey flocks, but its validity is
debated.

CONDITION   **SUDDEN DEATH SYNDROME**
Species   **Chickens: broilers and layers**

Clinical aspects   Broilers: individual well grown birds are found dead, sporadically, often on their back, 65-70% males.

Layers on deep litter: only hens affected, and usually only those in good condition and in full lay. Sporadic condition which has been associated with aggregation behaviour. Losses more common in autumn and winter.

Age range   Broilers - from 7 days onwards. Layers - from point of lay onwards.

Causal factors   Well grown birds subject to stress appear to be at risk. Behaviourial aspects are probably important in both conditions.

Diagnosis   A very healthy bird with few lesions except oedema and congestion of the lungs, ample food in crop and gizzard, and congestion of the ova in the case of hens, is the usual autopsy. History, epidemiology, and flock observations are helpful.

Immediate action   Remove carcases.

Economic importance   A significant cause of continuous broiler loss, in layers it can be a problem in heavy strains.

Welfare   Since the death is sudden the birds probably suffer only limited discomfort.

Footnote   Aortic rupture in young rapidly growing turkeys is a recognised cause of sudden death in that

species, which is easily diagnosed at autopsy.
Heat stress can cause heavy mortality in well
grown broilers in hot summer weather when
ventilation systems may prove inadequate.

Index